Newton & Descartes's Skate Park Adventure

MATH MUSICALS

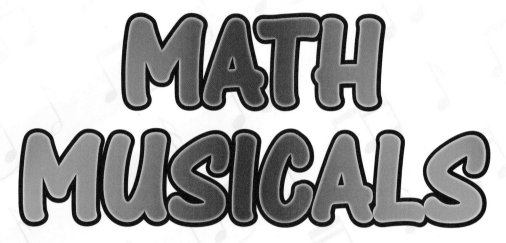

www.MathMusicals.com

Volume 1

by
Anne Lazo
and
Michael Wiskar

BIG IDEAS LEARNING®

Erie, Pennsylvania
BigIdeasLearning.com

Newton   *Descartes*

Contents

The authors of *Big Ideas Math* understand the importance of storytelling and music in learning. Storytelling puts math in real-life contexts, and music helps children remember. In collaboration with storyteller Anne Lazo and musician Michael Wiskar, the authors of *Big Ideas Math* are excited to offer *Math Musicals*—engaging musicals that bring mathematics to life!

In this fictional world, students can interact with many different characters, including Newton and Descartes! Students will often wonder—What adventures will Newton and Descartes have today? What will they learn? What will they do next? Using *Math Musicals* will not only engage students in the mathematics, but will help improve their literacy skills as well.

Math and literacy skills are critically important in this world. Bringing math and literacy together through music is the perfect way to instill a lifelong love of learning, encouraging young minds to love math. At Big Ideas Learning, we believe providing the opportunity for students to interact with and learn the math through storytelling and music is just one more way to make learning visible.

About the Authors

Anne Lazo is an international performing arts educator, theatrical director, and author and composer of children's musical educational books.

Anne graduated from the University of Southern California and has over 20 years experience teaching and managing inspirational arts programs and productions for students, as well as professional actors, musicians, and dancers.

Anne worked in Japan for 15 years as an accomplished musical theater director, acting coach, choreographer, script writer, director of educational programs, and motivational speaker for arts advocacy. She has worked for Disney's World of English (World Family English), NHK Television and Radio, and owned her own production company, where she produced and directed musicals for schools and companies. In the United States, Anne was the Director of Education for the New West Symphony, where she developed and implemented many community outreach programs for all ages throughout the Los Angeles area.

Anne currently lives with her husband in Spain and is the Creator and Founder of My Child and Me - English©. Anne and her theatrical team of actors and teachers provide ESL educational shows, musical storybooks, and classes for young children.

Michael (Mike) Wiskar is a Leo-nominated composer, songwriter, independent music producer, vocalist, and multi-instrumentalist.

A graduate of the Music Program at MacEwan University as well as Langara's Digital Music Production Program, Mike began his music career in his teens as a performer in rock, country, and jazz bands before working as a songwriter for various artists in Quebec. There he worked closely with Myles Goodwyn of legendary Canadian rock band April Wine, among other top-level songwriters and producers.

Mike has since written and produced songs and music scores for theater, short films, full length and series documentaries, choirs, contemporary artists, and numerous television series (Operation Vacation, Psycho Kitty, Men's Fashion Insider, Kitty 911, and Favorite Places, to name just a few).

Math Musicals Stories and Topics

Grade K

Newton & Descartes's Coolest, Rockin' Day Ever

- *Fish Crackers* (Counting to 10)
- *Cool Cats & Rockin' Dogs* (Partner Numbers to 5)
- *Big Bulldog Bob* (Counting to 20)
- *JAM Session* (Naming Two- and Three-Dimensional Shapes)
- *Best Friends (I Get You and You Get Me)*

JAM Session

100 Waves

Grade 1

Newton & Descartes's Day at the Beach

- *Seashells* (Addition to 10)
- *Cora's Home* (Subtraction to 10)
- *100 Waves* (Counting to 100 by 10s)
- *Racing the Clock* (Telling Time)
- *Best Friends (I Get You and You Get Me)*

Grade 2

Newton & Descartes's Four-Legged Fun

- *Four-Legged Games* (Two-Digit Addition)
- *Newton & Descartes's Turn* (Two-Digit Addition and Subtraction)
- *Pet Airlines* (Counting to 1,000 by 100s)
- *At the Airport* (Money)
- *Best Friends (I Get You and You Get Me)*

At the Airport

Grade 3
Newton & Descartes's Night in Madrid

- *Tiny Tapas* (Multiplying by 3)
- *Flamenco Mice* (2s, 3s, 4s, and 5s Multiplication Facts)
- *Where is the Park?* (Fractions)
- *Fútbol Fun* (Perimeter)
- *Best Friends (I Get You and You Get Me)*

Fútbol Fun

Newton Leads the Way

Grade 4
Newton & Descartes's Skate Park Adventure

- *Jump Start* (Estimating Sums and Products)
- *The Amphibian* (Divisibility Rules for 2, 3, and 5)
- *The Catwalk* (Adding and Subtracting Fractions with Like Denominators)
- *Newton Leads the Way* (Types of Measurement)
- *Best Friends (I Get You and You Get Me)*

Grade 5
Newton & Descartes's Pet Center Adventure

- *Happy Paws Club* (The Language of Decimals)
- *The Cat's Eye Club* (Multiplying Decimals and Whole Numbers)
- *Bark Like a Dog* (Adding Fractions with Unlike Denominators)
- *Friends* (Volume of a Rectangular Prism)
- *Best Friends (I Get You and You Get Me)*

Friends

Accompanying Products

Visit **www.MathMusicals.com** for the following products.

Songs

Math Musicals songs are available for use throughout the year. Play the songs from *Math Musicals* to not only help students learn concepts, but to help them review all year long as well!

Animated Videos

The *Math Musicals* songs and storybooks come to life in these engaging animations filled with vivid and fun images that your students are sure to enjoy. Join Newton and Descartes on adventures around the world! From rocking out at a jam party to flying high in an airplane, Newton and Descartes engage students in mathematics in fun and innovative ways!

Sheet Music

Each *Math Musical* song comes with its own sheet music. Students and teachers can engage in learning by playing along with songs from *Math Musicals* on a piano or a guitar! Teachers have the ability to truly mix the teaching of math and music with this easy-to-read sheet music.

Differentiated Rich Math Tasks

Differentiated Rich Math Tasks are a fun and intriguing way for students to interact with mathematics. These leveled worksheets challenge students to stretch their knowledge and dig deeper! Each task has three different levels so teachers can easily differentiate to meet the needs of every student.

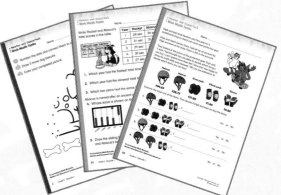

Puppets and Plush Toys

You can use the Newton and Descartes puppets and plush toys to make learning fun and to help students talk about math!

Newton & Descartes's Skate Park Adventure

by
Anne Lazo and Michael Wiskar

Jump Start (Estimating Sums and Products)
The Amphibian (Divisibility Rules for 2, 3, and 5)
The Catwalk (Adding and Subtracting Fractions with Like Denominators)
Newton Leads the Way (Types of Measurement)

JUMP START

by Anne Lazo

Nothing is better than going on a weekend trip! Miss Polly took her cat, Descartes, and her dog, Newton, to the big city. She was visiting her younger brother, Gram, who lives on the 20th floor of a high-rise apartment building. He has two cats, Digit and Axel. They were very excited that Newton and Descartes were visiting and could not wait to show them their favorite place in the city.

While Miss Polly and her brother were catching up in the kitchen, all four pets snuck out of the apartment to explore the city. Axel and Digit grabbed their skateboards and gear on the way out. They are skateboard enthusiasts and neither of them go anywhere without their helmets, protective gear, and boards.

As they waited by the elevator to go down to the street, Digit pulled out her mobile cat phone from her collar and put it over her ears. "Don't worry, we will be there soon," she said.

"Be where soon?" asked Descartes.

"We thought you both might like to go with us to the skate park and watch us practice. Then we can explore the city streets. Does that sound like a plan?" asked Digit as she proceeded to call her friend, Scooter.

"Hello, Scooter? Where are you now? Oh good. I don't know. It will take us about **8 minutes** to walk to the subway station and then about another **19 minutes** or so on the subway to the skate park exit. Anyway, we'll be down there in **about 30 minutes**. Is that okay? Great, see you there!"

Newton and Descartes looked at Digit in amazement. They had never seen a mobile phone made for cats before. They did not know that Digit and Axel were so crazy about skateboarding either!

"We've never seen a mobile phone for cats," meowed Descartes, somewhat skeptical.

"Gram only lets us go outside if we take one with us. The phone can track our location so that he always knows where we are. There is even a special button that beeps when he calls us for dinner!" Axel explained.

"I'm not allowed to roam around the neighborhood alone," said Newton. "Miss Polly has to take me for a walk everyday on a leash."

"That's the great thing about cats, we can come and go when we want to!" Axel said happily, as he looked forward to leaving the house again for the day. Descartes thought it might be interesting to own a mobile phone and wanted to show Miss Polly when they got back. Newton was a bit concerned about exploring the city with two funny cats that carry mobile phones and ride skateboards, but enjoyed the thought of being able to go wherever he wanted without a leash!

Digit gave Descartes Axel's mobile phone to use during their stay. She could see that Descartes was a bit clumsy with it, so she showed him how to use it while they rode down the elevator to the ground floor.

"Just put it over your ears like this. Push the number six and it will ring me. If you push the number seven, it will ring Gram. Or, we can set it to meow mode and you can meow six times and it will connect to me or meow seven times and it will dial Gram directly. Got it?" Digit explained.

"I guess so," said Descartes, with a bit of trepidation as the elevator doors opened and they exited onto the street. Newton giggled to himself at the sight of Descartes wearing the silly mobile phone on his head.

As the four pets arrived at the subway, Digit pulled out her subway card to pay for the tickets for the four of them. "I will buy the tickets for everyone. I am pretty sure that I have enough on this card to cover us. Let's see, how much do we need?" she asked.

"Each ticket is **$2.75**," replied Axel. "That's 2 dollars and 75 cents."

"That's about **$3. $3 × 4 is $12.** I am guessing that I have about $12 left on this card," meowed Digit happily. Sure enough, she put the card into the machine and the amount left was just enough to pay for everyone.

"Thank you very much, Digit!" meowed Descartes. "Newton and I will pay for the return tickets."

"No problem," replied Digit. "Axel and I are super happy that you are visiting and can come with us to the skate park today!"

Calling My Squad!

by Michael Wiskar

Calling my squad!
We're going to the skate park.
Don't want to be late.
We're going to skate hard.
Calling my squad!
Going to show them what we're made of.
Calling my squad!

Calling my squad!
We're going to the skate park.
Don't want to be late.
We're going to skate hard.
Calling my squad!
Going to show them what we're made of.
Calling my squad!

But first we've got to walk to the subway.
It's about 8 minutes then we're on our way.
Then it takes about another 19 by train.
We've got to **estimate**,
So we won't be late.

8 minutes is about 10 minutes.
19 minutes is about 20 minutes.
10 plus 20 is 30 minutes.

Calling my squad!
We're going to the skate park.
Don't want to be late.
We're going to skate hard.
Calling my squad!
Going to show them what we're made of.
Calling my squad!

And now we've got to figure out train fare.
$2.75 each to get us there.
And four of us have got to ride the train.
We've got to **estimate**,
So we can take the train.

$2.75 is about 3 dollars.
4 times 3 is 12 dollars.
It's about 12 dollars to ride that train.

Calling my squad!
We're going to the skate park.
Don't want to be late.
We're going to skate hard.
Calling my squad!
Going to show them what we're made of.
Calling my squad!

Calling my squad!
We're going to the skate park.
Don't want to be late.
We're going to skate hard.
Calling my squad!
Going to show them what we're made of.
Calling my squad!

Calling my squad!
Digit! Axel! Newton! Descartes!

Calling my squad!

They made it to the park in **about 30 minutes**, just as Digit predicted. She was so pleased that everything she **estimated** so far was working out perfectly. She was happy that she could pay for the subway tickets and that they made it on time to meet Scooter. He was looking out for them when they arrived.

"Hey, it's great to meet you both! Have you ever been to a skate park before?" Scooter asked Newton and Descartes.

"It's great to meet you too!" barked Newton enthusiastically. "This is our first time, it's really cool! Wow, that looks like a really steep ramp!"

"Yes, that's called a half-pipe and it is **about 6 feet tall**," said Scooter.

"Hey, that's how tall Gram is!" chuckled Axel.

"Let's show them how it's done, Digit!" meowed Scooter as he took a tight, fast curve and rolled into a launch ramp.

"How did you like that big air flip that Scooter just did, Descartes?" meowed Axel.

"Okay, here I go!" Digit announced as she winked at Newton and Descartes and took off at a glorious pace down to the opposite side of the skate park.

"Oh, and there goes Digit … awesome air kick flip, Digit! Now Scooter is riding the rails! Well, I am going to take off and join the fun!" meowed Axel as he grabbed his skateboard and tightened his helmet.

Newton and Descartes were speechless. They could not imagine being able to do tricks like that and felt nervous just seeing their friends in action. Digit and Axel must have been practicing for many years to be so skillful, thought Descartes as he looked on. "This is definitely a sport that you have to practice a long time before you can safely use the skate park!" he declared.

"Yes, you have to start practicing on the street first until you get the hang of it!" replied Newton.

As Newton and Descartes were admiring the skateboarders, the wind picked up significantly. Suddenly, they heard a loud sound from the sky getting closer and closer. Just then, a helicopter came and hovered above the skate park. The helicopter door opened and a cat dressed in fancy skateboard gear was lowered down slowly to the ground.

"Who is that? What is going on?" meowed Descartes, very surprised.

Axel rode over to Newton and Descartes and told them not to worry because the greatest skateboarding cat of all time just landed! "I can't believe she came today! We only see her once or twice a year!" Axel proclaimed with joy.

"Who is she and where is she from?" Newton asked, barking loudly. He could hardly hear what Axel was saying because of the helicopter noise.

"No one knows her real name or where she is from. We just call her Ultra! She travels to skate parks all over the world and teaches cats, like us, the latest tricks! Ultra is amazing!" Axel continued. "Check out her skateboard. Have you ever seen anything like that before?"

Newton and Descartes looked at Ultra's skateboard and sure enough, it looked quite magical!

THE AMPHIBIAN

by Anne Lazo

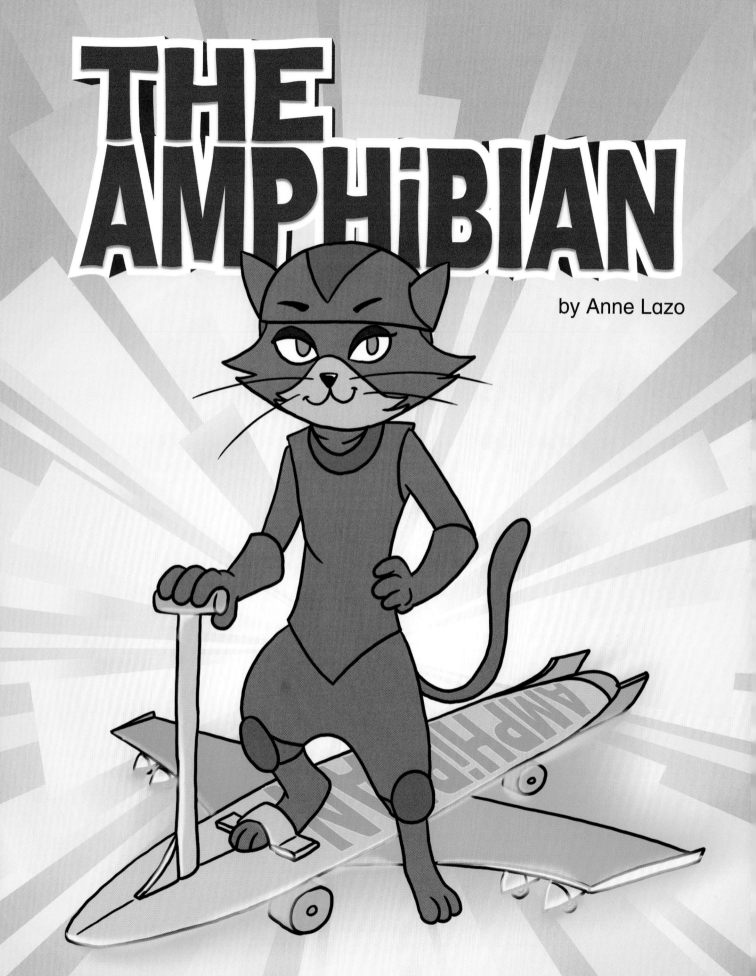

"Gather around, friends," Ultra meowed as she motioned for all of the cats to come to the middle of the skate park where she landed. "It's good to see all of you here today! I've come to show you the latest and greatest skateboard, the *Amphibian*! It has incredible, new technology. This baby can roll on land, fly in the sky, and glide across water."

"I'll show you how it works and hand out a bunch of them for you to try! The *Amphibian* is easy to use and it can get you around the city quickly. It will take you where you want to go, regardless of the type of land form," she continued.

16

Digit and Axel looked very interested in trying the new board. Newton and Descartes did not think it was possible for them to use it since they did not have any experience using a normal skateboard. Newton knew that it was made for cats and felt left out by the whole situation in general.

Ultra sensed that the only dog in the park was feeling excluded, so she walked over to him and asked him his name, "Hi there. And who is this nice dog hanging out with these cool cats?"

"My name is Newton," he replied.

"Well, Newton, I have something special for you. Check this out... My company made a prototype of this board especially for dogs! This one is your size. Would you like to try it?" Ultra asked as she uncovered the last board and handed it to Newton.

"Gee, it's not even heavy," barked Newton, as he held the board, feeling quite courageous.

Descartes gave Newton the evil eye and said, "Don't even think about it, buddy!"

Ultra appealed to the crowd of skateboarding cats, "Try this fantastic, new skateboard for a day and then come back and let me know if you like it. I'll let you keep the board if you complete the trial run. Be careful though! Please pay close attention to my demonstration and stay here and practice before you venture off in the city. We don't want anyone to get hurt! Now... How many of you would like to practice?"

All of the skateboarding cats learned how to hang onto the bar, strap their legs in, and move by leaning forward, backward, and sideways. They could even maneuver the board from their mobile phones. Ultra arrived by helicopter on purpose to show the cats what a nuisance it was trying to get around that way, compared to using the *Amphibian*.

"You can get through tight spaces and land anywhere. Plus, it's very quiet," Ultra boasted.

All of the skateboarding cats were enthusiastic about trying the *Amphibian*. They could not believe their luck!

Newton and Descartes, on the other hand, were disputing whether or not to participate in this, potentially absurd and dangerous, trial run. Descartes was completely against it. Newton felt ready for the challenge.

"Come on, Descartes. We have to try this! It will be fun!" barked Newton.

There were 15 pets gathered around Ultra now. "I'd like to divide all of you into groups with the same number of pets in each group," she declared. "Now... How many of you should be in each group?" asked Ultra.

"I can help with that," Descartes chimed in. "To break 15 pets into equal groups, you can have 3 groups or 5 groups, but you can't have 2 groups," said Descartes confidently.

"How do you know that?" asked Axel. Descartes was happy to show off his math ability. So, he proceeded to give all of the pets a lesson on divisibility. "Well, it goes like this," meowed Descartes, "**15 is not divisible by 2 because the last digit is an odd number.**"

"You could have **3 groups of 5 because 15 is divisible by 3**," purred Descartes. "I know that because **the sum of the digits 1 and 5 is 6 and 6 is divisible by 3**."

"Or, you could have **5 groups of 3 because 15 is divisible by 5**. I know that **because the last digit of 15 is a 5**."

All of the pets were impressed with Descartes's knowledge of mathematics.

"Well done!" Ultra praised Descartes. "I like the option of three groups with five in each group. That way, one group can ride their boards to the west side of the city, the second group can ride to the north side and the third group can test the *Amphibian* board on the south side. Then, everyone can make their way back here, to the east side! It works out perfectly!"

Here Comes Ultra (Divisibility Song)

by Michael Wiskar

She came from outer space
   On a magic skateboard.
And no one knows her name
   So we call her Ultra.
Everybody wants to be her friend,
Flying on their new *Amphibian*!

Here comes Ultra!
Here comes Ultra!

Here we go.

Ultra! Ultra! Ultra!

I see there are 15 of you.
And I want to divide you into groups,
With the same number in each group.
But I can't divide by 2
   because 15 is not divisible by 2.

As Descartes figured out.

So, we can have 3 groups of 5
   because 15 is divisible by 5.
We can have 5 groups of 3
   because 15 is divisible by 3.

Everybody understand?
We've got 3 groups of 5.
Now get on your boards and ride, ride.
Get on your boards and ride! Ride!

Here comes Ultra!
Get on your boards and ride! Ride!
Here comes Ultra.
Get on your boards and ride! Ride!
Here comes Ultra.
Get on your boards, get on your boards and ride!
   Ride!
Here comes Ultra.

Get on your boards and ride, ride, ride!

She came from outer space
   On a magic skateboard.

All of a sudden, Descartes realized that he and Newton should not be counted. So, he meowed, "But since Newton and I can't go, you need to find two more cats to join."

"Nonsense, Newton is the perfect leader for our first group of five going to the west side of the city. Digit, Axel, Scooter, and you must go with Newton. Lend him your mobile phone so that he can learn to navigate," Ultra suggested to Descartes.

"This is a great opportunity, Newton! Because of you, other dogs will see how much fun all of you are having and want to try the *Amphibian* board too! You won't have to wait in traffic or get stuck in crowds of people walking on the street. Just ride above them!" she exclaimed as she laughed with pride.

"I'm sorry, but Newton can't lead a group, miss, we are really not interested in participating in the trial!" Descartes whined.

"Newton and Descartes are just visiting us today. They came to watch us skate and then we were going to show them around," Digit explained to Ultra.

"Perfect! The great thing about the *Amphibian* is that anyone can ride it. You don't have to be good at skateboarding! It's the best way to tour the city! You are all going to have a great time. Don't worry, be happy! I'll be here when you get back, and believe me, you won't want your ride to end! Enjoy yourselves, west side group. I mean, west side squad!" Ultra said as she smiled, gave Newton and Descartes a big thumbs up, and sent them on their way.

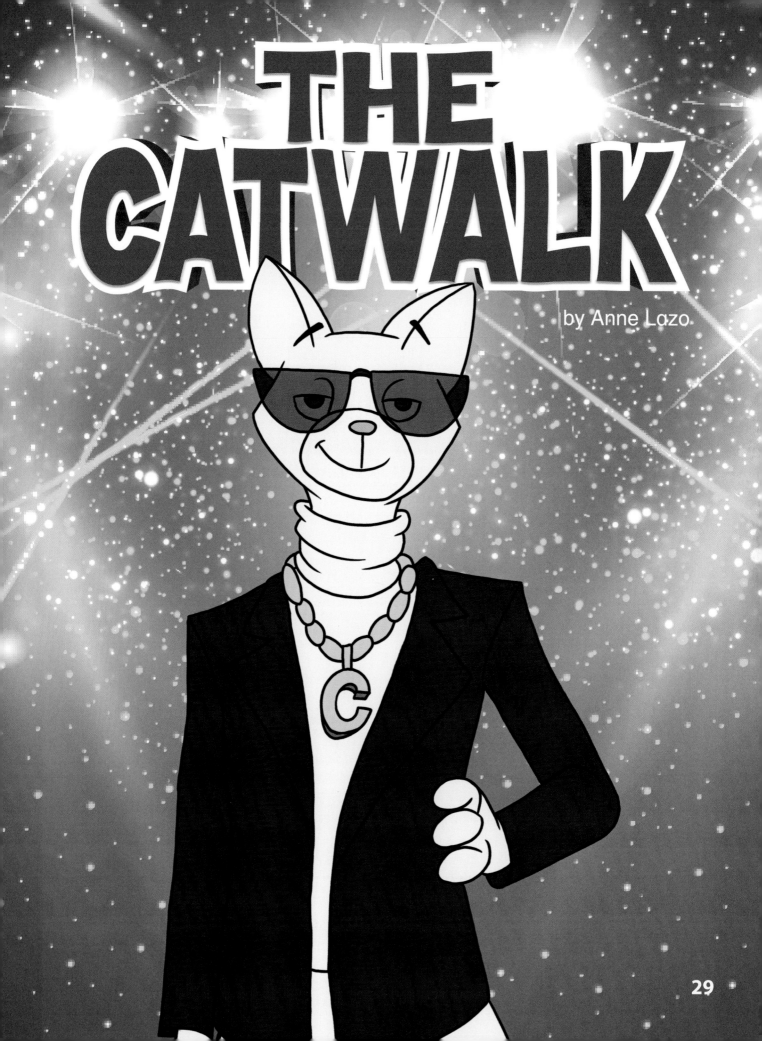

THE CATWALK

by Anne Lazo

"Where are we going?" asked Descartes, a bit confused as to how he ended up on an *Amphibian* skateboard, riding around the city streets.

"Well, we didn't expect to show you around the city this way, but this is awesome!" meowed Digit. "Hey, I thought Newton was the leader? He has to learn how to navigate!" Digit pointed out as they all slowed down to take a sharp turn. They were heading in the direction of the shopping district.

"That's okay, Digit. I am just trying to stay on this thing, never mind worrying about which way to go! It's better if you and Axel lead the way!" Newton barked, adamant that he was not comfortable being the leader in a city that he was not familiar with. Newton wondered why Ultra wanted him to lead the squad when he had no experience on a skateboard.

As the fabulous west side squad leaned forward on their boards, they tried to pick up speed on the ground. Digit wanted to try some tricks on and off the curb and up and down some stairs, but she was finding it difficult. "This thing doesn't really respond the same way as a regular board. It's hard to do tricks on it," she said.

"I think it's made more for, traveling than doing tricks," Axel agreed.

"Let's see if this baby can really fly," Scooter meowed excitedly. "Watch me get some air!" Scooter continued as he motioned for the squad to follow him. He told them to press the airlift button and hang on tight!

"WHOA," called out Newton.

"YIPPIE!" yelled Axel! The *Amphibian* was definitely a fun and fast board. It turned out to be pretty easy for the squad to ride on land and in the sky! Digit, Axel, Scooter, Newton, and Descartes were able to get enough height to fly over animals, cars, and even stores. They caused quite a stir on the ground below as other pets looked at the five of them and marveled at the sight. Dogs and cats were excited to see them fly by like superheroes. The fabulous west side squad waved to the pets below, while beaming with delight.

The International Pet Fashion Week was taking place. Famous designer pets from all over the world gathered to show off their latest pet wear. Some of the designers taking part this year were Dogatella Versace, Hugo Purr, and Jean-Paul Catier.

Monsieur Catier specializes in avant-garde cat wear. He was waiting nervously in his fashion tent for his last supermodel to arrive. His runway show had already started and the model cats were showing off his futuristic clothes on the high-tech catwalk.

"Where is that supermodel?" meowed Monsieur Catier angrily. "If he does not show up right now, we are doomed!"

As the fabulous west side squad approached the start of the main avenue, they could see that they were flying over a huge fashion event. They quickly made a swift turn into an alleyway to avoid flying over the fashion crowd. Descartes misjudged his turn and accidentally bumped into a sign post, which caused his board to go flying in the other direction. He landed right in front of Monsieur Catier's fashion tent.

"AAGH! Get him dressed and out on that catwalk! Quickly, quickly!" Monsieur Catier yelled to his dressmaker as he scolded Descartes for being late.

"This cat is smaller than we expected, sir," the dressmaker complained.

"Make it work!" Monsieur Catier demanded.

Before Descartes could explain that he was not a model, the dressmaker grabbed the *Amphibian* board from his paws and started pinning a futuristic-looking garment on him. The outfit was way too wide for Descartes, just as the dressmaker suspected. He had to get it down to $\frac{1}{4}$ **meter in width** to fit Descartes perfectly.

"**This outfit is $\frac{5}{8}$ meters in width, if I cut off $\frac{3}{8}$ meters then it will be $\frac{2}{8}$, or $\frac{1}{4}$, meters wide.** Then it will fit him perfectly," meowed the dressmaker in a panicked and hurried voice.

"Alright, make it happen. I have to have this 'look' just right for the big finale! Oh, and I like that skateboard thing. It matches the entire feel of the collection. Make sure he goes out with that too," Monsieur Catier asserted, intrigued by the board's absolute splendor!

Descartes was beside himself with confusion and disbelief. What was happening? And how did he get himself into this situation?

Meanwhile, back in the alleyway, Digit noticed that Descartes was missing. "We have to go back!" she called out. Everyone stopped and turned around. "I think we lost Descartes. He must have fallen down near the fashion tents!"

"What?" Newton replied, shocked. "Poor Descartes! I hope he is not hurt! It's my fault. I should not have pushed him to ride along with us!"

"Don't worry, I'm sure he is fine. Let's get off of our boards, walk over there, and look for him," Digit said while she tried to calm Newton's nerves.

"It's alright, I'll get him, Digit. You stay here and wait for us," Newton said, feeling responsible for the whole incident.

After looking all over for Descartes, Newton finally found him in Catier's tent. When Monsieur Catier saw Newton enter with his *Amphibian* skateboard, he thought that he was another model that was lost.

"Another model with a superb skateboard? How wonderful! We must dress him too. I want the fashion world to see my fabulous sci-fi designs with our skateboarding cat and DOG!" He pulled Newton over to the dressmaker to get him dressed immediately.

Newton was relieved to see that Descartes was not hurt, but could not help laughing when he saw him all dressed up in a futuristic cat outfit. "Hey, Descartes, I think that outfit suits you, my friend!"

36

"You are not funny at all, Newton! It looks like you got roped into being in the show, too!" Descartes meowed back, completely irritated by the whole fashion tragedy!

"This is fabulous beyond my belief! They are both so good looking! Yes, we must have a dog and cat come out for the finale with those fabulous skateboards!" Monsieur Catier declared with glee!

"Good looking? What is he talking about?" Descartes said, looking at Newton in disbelief.

"Yes, well, I always thought I had a special, unique look," Newton replied with a cheeky grin.

"But, sir, there is no more time. We don't dress dogs! He is much bigger than a cat! You make all of your designs for cats, not dogs! I don't have anything quite that big," the dressmaker kept repeating while he sweated profusely.

"Use the leftover material from this cat's outfit and add more from the pile of material over there. I will help you," Monsieur Catier said. He was looking for material that he could mix and match to come up with a new outfit for Newton.

They measured out $\frac{1}{4}$ meter of the old material from Descartes's outfit and added to it another $\frac{3}{4}$ meter of material, which made a total of $\frac{4}{4}$ meter, or 1 meter. They had a new fabulous look for Newton in no time. It looked even better than Descartes's design!

"This is better than I ever imagined!" purred Monsieur Catier, extremely proud of his final masterpieces! "Go! Go! Go! Get out there with your boards!" he shouted as he pushed Newton and Descartes onto the catwalk.

All Dressed Up

by Michael Wiskar

Walk. Walk. Walk. Walk. Walk. Walk.
Walk. Walk. Walk. Looking sharp!
When the time to hit the stage arrives,
And you've got to get your clothes fit right to size,
And you look so good you can't believe your eyes,
You make the crowd go crazy with your funky style.

Yes, we're all dressed up,
And there's nowhere we can't go.
Yes, we're all dressed up, dressed up.
That's right, looking sharp!
Yes, we're all dressed up,
And there's nowhere we can't go.
Yes, we're all dressed up, dressed up.
That's right, looking sharp!

Now you've got to figure out your size.
And **they've got 5/8 meter**
    Much too wide!
They've got to **subtract 3/8** so it's tight.
That's tight! **5/8 minus 3/8 is 2/8.**
That's right!

Now you're all dressed up,
And there's nowhere we can't go.
Yes, we're all dressed up, dressed up.
That's right, looking sharp!
Yes, we're all dressed up,
And there's nowhere we can't go.
Yes, we're all dressed up, dressed up.
That's right, looking sharp!

Hey, hey, hey, Newton.
Got to make a new outfit for you.
They've got **1/4 meter** from
    Descartes's new outfit,
We need 3/4 more for you.
1/4 plus 3/4 is 4 quarters.
That's true!
4 quarters is 1 meter.

Now let's do what we've got to do.
Now let's do what we've got to do.
Now let's do what we've got to do.
Now let's do what we've got to do!

Now, you're all dressed up,
And there's nowhere we can't go.
Yes, we're all dressed up, dressed up.
That's right, looking sharp!
Yes, we're all dressed up,
And there's nowhere we can't go.
Yes, we're all dressed up, dressed up.
That's right, looking sharp!

Now you're all dressed up,
And there's nowhere we can't go.
Yes, we're all dressed up, dressed up.
That's right, looking sharp!
Yes, we're all dressed up,
And there's nowhere we can't go.
Yes, we're all dressed up, dressed up.
That's right, looking sharp!

Now you're all dressed up.
Looking sharp!

Newton and Descartes felt timid and scared at first. Not knowing what to do, they looked at each other and just started laughing. They decided to have fun and make goofy poses with their skateboards in their paws. They strutted down the catwalk with a playful confidence.

As the sound of the electric beat in the background grew in intensity, they threw down their *Amphibian* boards, hit the airlift button, and took off to the explosive applause of the audience!

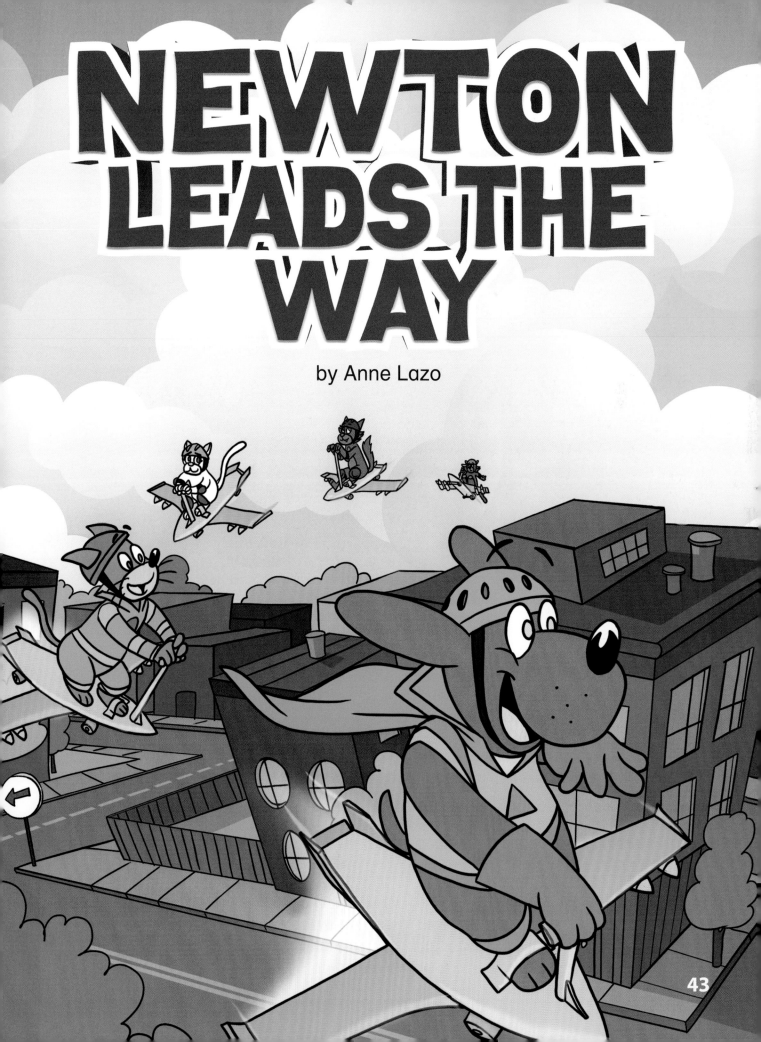

NEWTON LEADS THE WAY

by Anne Lazo

Newton grabbed the mobile phone and quickly pushed "#6" to call Digit. "Hey, this thing really works," he thought. Newton told his squad to come over to Monsieur Catier's tent right away. They just had to see what they were up to!

Digit, Axel, and Scooter rushed over to the scene. All three of them were wondering what all the fuss was about at the designer's tent. Suddenly, they saw Newton and Descartes standing with Monsieur Catier surrounded by photographers.

44

"What are they doing?" Axel meowed in disbelief!

Newton motioned for his friends to come over. "Well, I found Descartes!" he said. "It seems he always had a secret longing to become a famous model. I just couldn't let him do it alone, so I joined in the fun!" Newton barked in jest to his friends.

"I can't believe it. I've lived here all of my life and I've never even been able to get a ticket to one of these events!" Digit proclaimed, very impressed. "I don't know how you did this!"

"The *Amphibian*!" both Newton and Descartes called out happily. "It made us look like we knew what we were doing and gave us confidence!" Newton admitted.

"You do know what you are doing, Newton!" Digit confirmed with a grin. "Why don't you get us out of here now and we'll continue on your spectacular tour!"

The fabulous west side squad took to the air again and was in the flow with Newton in the lead. He wanted to explore different parts of the city that were off the beaten path. Digit and Axel were excited because everywhere he wanted to go was new for them, too. They rode up and down the west side of the city and checked out the Animal Art Museum, the Garden of Animal Statues, and a really cool antique game arcade center.

They tried to ride the handrails and ledges around the buildings, but the *Amphibian* was too unsteady for street stunts. After a few hours of fun, Newton suggested they stop and have a picnic in the park. Axel agreed that they could all use a break and suggested a deli, not too far away, that had a great selection of take-out, gourmet pet food.

46

"How are you all doing today?" the dog behind the counter asked the squad. "Hey, those are fancy boards you have there. Are you from out of town or something?" the deli dog wondered as he looked at Newton and Descartes's strange outfits.

"Yes, those two are definitely from out of town," Axel answered jokingly.

"If you are visiting, then you have to try the most famous food in the city! We have delicious purritos, collieflower, bark-B-Q sandwiches, or how about a corn dog!" the deli dog suggested.

"I think I'd just like to have the **1-pound gourmet bone with 8 ounces of boiled potato salad**," Newton said, as he sniffed around the counter.

0 1 2 3 4 5 6 7 8 9 10 11 12

12 in. = 1 ft

6-inch salmon sandwich

1-cup fish soup

1 lb = 16 oz

1 cup = 16 fl oz

8-oz coleslaw

square of lasagna

4 in

4 in

1 quart of salad

Each of the cats stepped up to the deli counter to order.

Digit purred, "I'll have a **6-inch** salmon sandwich."

Axel said, "I don't feel like a sandwich today. I'd like **1 cup** of the fish soup with **8 ounces** of coleslaw."

Scooter ordered next. "I'd like a **4-inch square piece** of the tuna lasagna."

Last, it was Descartes's turn to order. "I want to stick with vegetarian items. I'd like a salad in one of those big **quart bowls**. And can you throw some fish crackers on the top?"

However You Like It! (The Measurement Song)

by Michael Wiskar

(Descartes) Hey Newton.

(Newton) Yeah, Descartes.

(Descartes) Look at this menu.
You can get *anything*. Any size!

(Newton) *Any size*???

(Descartes) Yeah, you just ask him.

(Newton) Ok!

(Descartes) Order up!

How big? How small? However you like it.
You make the call. However you like it.
We've got it all. However you like it.
Now, I've just got to know if it's to stay or
it's to go.

So what'll it be?

You want a **1-pound gourmet bone**.
(Newton) And **8 ounces of potato salad**
please (on the side).
And for you a **6-inch salmon sandwich**.
And for you **1 cup of fish soup** (soup,
soup).
With **8 ounces of coleslaw** (coleslaw).
And a **4-inch square** of tuna lasagna.

And the deli dog says:

How big? How small? However you like it.
You make the call. However you like it.
We've got it all. However you like it.
Now, I've just got to know if it's to stay or
it's to go.

So here's how it works.
Anything like meat or bones (coleslaw).

I sell them by **weight**.
That's **pounds and ounces**.
Anything that's a **liquid**, like a soup (soup, soup),
Is measured in cups. Or quarts even!
Now, something like a **sandwich**,
that's measured in inches.
Or feet, if you're really hungry!
Got it? Let's eat!

And the deli dog says:

How big? How small? However you like it.
You make the call. However you like it.
We've got it all. However you like it.
Now, I've just got to know if it's to stay or it's to go.

How big? How small? However you like it.
You make the call. However you like it.
We've got it all. However you like it.
Now, I've just got to know if it's to stay or it's to go.

How big? How small? However you like it.
You make the call. However you like it.
We've got it all. However you like it.
Now, I've just got to know if it's to stay or it's to go.

Thanks, fellas!

(Newton) Um, um excuse me, I never did get a
potato salad. I didn't, excuse me, excuse me
. . . I . . .

Holding their delicious deli food with one paw and hanging on with the other, the squad hovered in the air once again and headed for the park. They found a perfect tree near a big pond to rest and eat. They thought about how the day ended up completely different than they expected. A day at the skate park turned into a whirlwind experience around the city.

"Hey, we haven't tried the board on water yet," Scooter meowed, as he nibbled his tuna lasagna in a hurry. Scooter was eager to try the *Amphibian* on the pond. He jumped on the board and rode over to the water. Then he hovered over the pond and slowly lowered the board, testing the water to make sure the board did not sink. "The *Amphibian* really floats on water too!" he called out to the others. "I'll try to glide it across the pond now."

Several ducks were watching Scooter as he flew and glided on the pond. They thought it was strange that a cat was trying to behave like a duck, moving on land, air, and water. They figured he was trying to take over their territory and attack them. The ducks flew at Scooter at a feverish pace to try and scare him away.

"Hey, a couple of ducks are attacking Scooter out there!" Digit meowed in alarm.

Newton thought quickly. He took off his designer wear, left his board behind, and ran into the pond. The ducks flew so hard into Scooter's board that they caused it to spin out of control. Scooter lost his balance and went flying into the water. Newton did not waste any time barking at the ducks as he pulled the board over to Scooter and dragged him back onto the grass.

The squad cheered as the two of them came back and sat down under the tree. "Gee, I guess those ducks were trying to tell us something. They don't want cats swimming in their pond!"

"That's true. We definitely ruffled their feathers!" Digit replied as the squad laughed.

"Actually, it felt good to run in the pond and swim without the *Amphibian* board. And without this hot outfit!" Newton said as he shook off the water.

"Let's go back to the skate park now. I think we've seen enough sights for one day," Descartes added as he took off his outfit, too. "Newton, do you think you can remember the way back to the skate park?" he asked.

"I'm pretty sure I can remember the way. This city is not as difficult to navigate as I thought it would be. I didn't even need the help of this mobile phone!" Newton barked while he wagged his tail.

"Good for you, Newton. You didn't need our help either! We all had a great time and it's been more fun than riding the subway everywhere!" Digit remarked, feeling happy that she saw new sights too.

So the fabulous west side squad got back on their *Amphibian* boards and airlifted one last time to the skate park.

"Well, Newton, you made it back safely with your squad! Did you all have a good time?" asked Ultra.

"He was a great leader, even though he is new here and never rode a skateboard before. It turns out that the *Amphibian* is very easy to use, even for beginners, just like you said!" Digit proclaimed.

"Yes, you could even say he was a model pet!" Axel said while the squad tried to hold in their laughter.

Digit looked out at the ramps in the skate park and put down the *Amphibian*. She grabbed her own board and said to Ultra, "You know, as much fun as we've had together, I think I like my skateboard better. The *Amphibian* is fast and convenient to use, but I can't do stunts on it or try different moves. I look forward to coming to this skate park every week and practicing with my board. I get better and better every time!"

"It is amazing how you can fly and glide on the water, and we did get a lot of attention. But, some animals didn't appreciate us flying over their home," Scooter added.

"Well, it opened up opportunities and gave me a lot of confidence, but I have to admit, I prefer running instead of using a board," Newton admitted.

Ultra was pleased to hear about all of their experiences on the *Amphibian* boards. As the helicopter started up again, she thanked them for participating in the trial run and put the boards back in her bag. Descartes felt that Ultra actually looked happy that none of them wanted to keep the *Amphibian*. Ultra bid farewell to the fabulous west side squad and the other squads that returned to the skate park earlier. She gave them another thumbs up and told them that she would be back soon!

Digit, Axel, Scooter, Newton, and Descartes stared at Ultra with wonder as she left. Once again, she taught them a lot. Suddenly, Axel's mobile phone beeped. "Oh, it looks like Gram is trying to call you," Newton said, as he gave Axel back his phone. "It must be time for dinner!"

Best Friends
(I Get You and You Get Me)

by Michael Wiskar

A wink, a nod, a sideways glance,
   a little elbow to the ribs.
We speak a secret language and
   no one else knows what it is.
You know me better than
   I even know myself.
It's a mutual connection
   between us and no one else.

You've got a friend,
Whenever you need a friend.
You've got a friend,
Whenever you need a friend.
I can't pretend
   there's anywhere else I'd rather be,
'Cause I get you and you get me.

And even when there's nothing to do
   we're never feeling bored.
We've had a thousand adventures
   and we'll have a thousand more.
You make me laugh so hard
   I feel like I could cry.
And we pick each other up
   when we're together,
We don't even have to try.

You've got a friend,
Whenever you need a friend.
You've got a friend,
Whenever you need a friend.
I can't pretend
   there's anywhere else I'd rather be,
'Cause you've got a friend in me.

'Cause I get you and you get me.
'Cause I get you and you get me.
We're a special kind of family.

And even if there's trouble brewing
    we always know just what we're doing,
'Cause I've always got your back
    and you've got mine.
And everything's fine!

You've got a friend,
Whenever you need a friend.
You've got a friend,
Whenever you need a friend.
I can't pretend
    there's anywhere else I'd rather be,
'Cause I get you and you get me.

Whenever you need a friend.
    ('Cause I get you and you get me.)
Whenever you need a friend.
I can't pretend
    there's anywhere else I'd rather be,
'Cause you've got a friend in me.

'Cause I get you and you get me.
    (Yeah, I get you!)
'Cause I get you and you get me.
    (I think you get me too.)
We're a special kind of family.

Let's sing it!

You've got a friend,
    ('Cause I get you and you get me.)
Whenever you need a friend.
You've got a friend,
    ('Cause I get you and you get me.)
Whenever you need a friend.
You've got a friend,
    ('Cause I get you and you get me.)
Whenever you need a friend.
We're a special kind of family.

You've got a friend,
    ('Cause I get you and you get me.)
Whenever you need a friend.
You've got a friend,
    ('Cause I get you and you get me.)
Whenever you need a friend.
You've got a friend,
    ('Cause I get you and you get me.)
Whenever you need a friend.
We're a special kind of family.

You've got a friend,
    ('Cause I get you and you get me.)
Whenever you need a friend.
You've got a friend,
    ('Cause I get you and you get me.)
Whenever you need a friend.
You've got a friend,
    ('Cause I get you and you get me.)
Whenever you need a friend.
We're a special kind of family.

Calling My Squad!

Words and Music by
Michael Wiskar

59

60

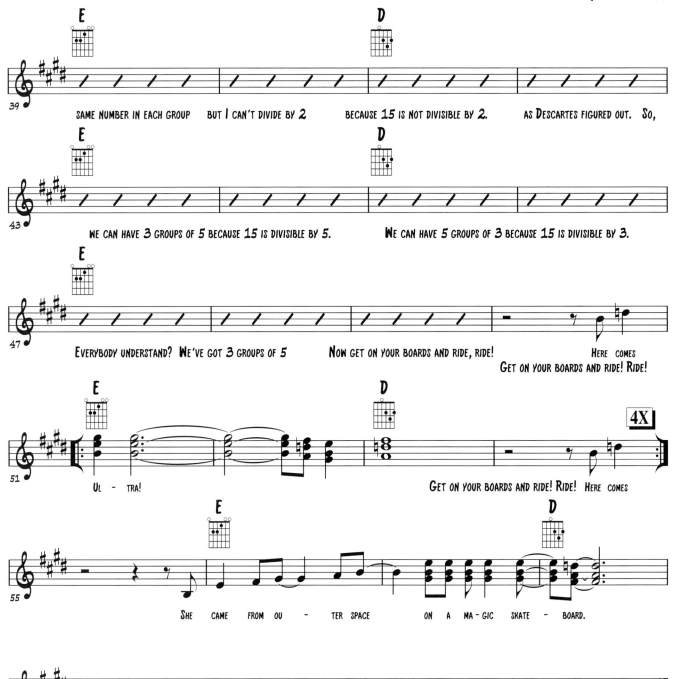

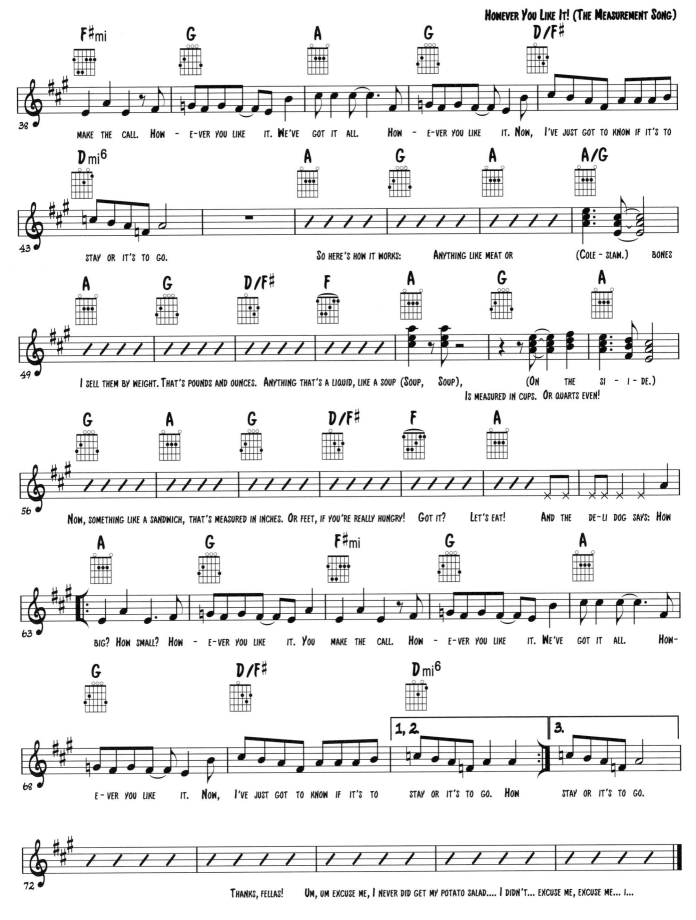

66

Best Friends
(I Get You and You Get Me)

Words and Music by
Michael Wiskar